FOXES

by Martha London

Cody Koala
An Imprint of Pop!
popbooksonline.com

abdobooks.com
Published by Pop!, a division of ABDO, PO Box 398166, Minneapolis, Minnesota 55439.

Printed in the United States of America, North Mankato, Minnesota

082020
012021

THIS BOOK CONTAINS RECYCLED MATERIALS

Cover Photo: Shutterstock Images
Interior Photos: Shutterstock Images, 1, 10–11, 12–13, 19 (top), 19 (bottom left), 19 (bottom right), 20; iStockphoto, 5 (top), 5 (bottom left), 5 (bottom right), 6, 9, 15, 16, 17, 21

Editors: Christine Ha and Brienna Rossiter
Series Designer: Sophie Geister-Jones

Library of Congress Control Number: 2019954984

Publisher's Cataloging-in-Publication Data
Names: London, Martha, author.
Title: Foxes / by Martha London
Description: Minneapolis, Minnesota : POP!, 2021 | Series: Underground animals | Includes online resources and index.
Identifiers: ISBN 9781532167614 (lib. bdg.) | ISBN 9781532168710 (ebook)
Subjects: LCSH: Foxes--Juvenile literature. | Foxes--Behavior--Juvenile literature. | Burrowing animals--Juvenile literature. | Underground areas--Juvenile literature.
Classification: DDC 599.775--dc23

Hello! My name is

Cody Koala

Pop open this book and you'll find QR codes like this one, loaded with information, so you can learn even more!

Scan this code* and others like it while you read, or visit the website below to make this book pop.

popbooksonline.com/foxes

*Scanning QR codes requires a web-enabled smart device with a QR code reader app and a camera.

Table of Contents

Chapter 1

Digging Dens

Foxes are **mammals**. They live on every **continent** except Antarctica. Some foxes live near forests and fields. Others live in deserts. Foxes even live in the Arctic.

Watch a video here!

All foxes make their homes underground. They dig dens. Each den has several exits. That way, the foxes can escape if a **predator** gets into the den.

Chapter 2

Bushy Tails

Foxes have **bushy** tails. Their teeth and claws are sharp. Some kinds of foxes are just 9 inches (23 cm) long. Others can be longer than 34 inches (86 cm).

Learn more here!

Foxes have thin legs. They have thick fur. Their fur can be red, white, gray, or black. The different colors help each fox blend in with its **habitat**.

A fox's fur has two layers. It protects the fox from rain, snow, and wind.

ear
snout
legs

A fox's tail is nearly as long as its body. The tail helps the fox stay balanced. It can also act as a blanket. The fox can wrap the tail over its nose to keep warm.

Chapter 3

On the Hunt

Foxes are **omnivores**. They eat both plants and animals. Foxes often look for food at night. They hunt small animals such as squirrels, mice, and rabbits.

Learn more here!

Most foxes hunt alone. Foxes have excellent hearing. They also see well in the dark. Foxes run fast to catch

their **prey**. Sometimes they even climb trees.

Red foxes can run up to 30 miles per hour (48 km/h).

Chapter 4

Safe and Warm

A fox's den has many rooms. Some rooms are used for storing food. Others are used for sleeping. Foxes also raise their young in their dens.

Complete an activity here!

Young foxes stay in the den for approximately one month. Then they begin to go outside. Eventually,

the foxes make dens of their own.

A mother fox piles leaves in its den to make a nest. The leaves keep baby foxes warm and dry.

Making Connections

Text-to-Self

Foxes can be many colors. What color would help a fox blend in where you live?

Text-to-Text

What books have you read about other mammals? What do those mammals have in common with foxes? How are they different?

Text-to-World

Foxes have thick fur. What other ways do animals stay warm in cold weather?

Glossary

bushy – thick and fluffy.

continent – one of the seven large landmasses on Earth.

habitat – the area where an animal normally lives.

mammal – a type of animal that has hair or fur and feeds milk to its young.

omnivore – an animal that eats both plants and other animals.

predator – an animal that hunts other animals for food.

prey – an animal that is hunted by other animals.

Index

Online Resources

popbooksonline.com

Thanks for reading this Cody Koala book!

Scan this code* and others like it in this book, or visit the website below to make this book pop!

popbooksonline.com/foxes

*Scanning QR codes requires a web-enabled smart device with a QR code reader app and a camera.